FI

HIGH ANGLE
RESCUE TECHNIQUES

Third Edition

FIELD GUIDE to Accompany
HIGH ANGLE
RESCUE TECHNIQUES
Third Edition

Tom Vines
Training Officer
Carbon County Sheriff's Search and Rescue
Red Lodge, Montana

Steve Hudson
President, Pigeon Mountain Industries, Inc.
Deputy Director, Walker County Emergency
Management
Lafayette, Georgia

With 172 illustrations

JONES & BARTLETT
LEARNING

World Headquarters
Jones & Bartlett Learning
5 Wall Street
Burlington, MA 01803
978-443-5000
info@jblearning.com
www.jblearning.com

Jones & Bartlett Learning books and products are available through most bookstores and online booksellers. To contact Jones & Bartlett Learning directly, call 800-832-0034, fax 978-443-8000, or visit our website, www.jblearning.com.

Substantial discounts on bulk quantities of Jones & Bartlett Learning publications are available to corporations, professional associations, and other qualified organizations. For details and specific discount information, contact the special sales department at Jones & Bartlett Learning via the above contact information or send an email to specialsales@jblearning.com.

Production Credits
Chief Executive Officer: Ty Field
President: James Homer
Chief Product Officer: Eduardo Moura
Executive Publisher: Kimberly Brophy
Executive Acquisitions Editor—Fire and Electrical: William Larkin
Vice President of Sales, Public Safety Group: Matthew Maniscalco
Director of Sales, Public Safety Group: Patricia Einstein
Production Editor: Tina Chen
Director of Marketing: Alisha Weisman
Manufacturing and Inventory Control Supervisor: Amy Bacus
Director of Photo Research and Permissions: Amy Wrynn
Printing and Binding: Courier Companies
Cover Printing: Courier Companies

ISBN: 978-1-284-04391-4

6048

Printed in the United States of America
17 16 15   10 9 8 7 6 5 4 3

Contributors

Michael V. Callahan MD, MSPH, DTM&H
Medical Director, Rescue Medicine
CIMIT/ Massachusetts General Hospital
Boston, Massachusetts

Jim Kovach
Rescue Instructor
Cuyahoga Valley Career Center
Brecksville, Ohio
Bowling Green State Fire School
Bowling Green, Ohio

Loui McCurley
Technical Specialist
Alpine Rescue Team, Colorado
VP Technical Marketing
Pigeon Mountain Industries, Georgia

Ken Phillips
Chief of Emergency Services,
National Park Service
Grand Canyon National Park, Arizona

Contents

WARNING!

This guide is not a complete instructional manual for rope rescue. It is designed as a field reference for rope technicians who have already received training. It is not intended as a substitute for training.

High angle rope skills cannot be learned simply by reading a book. Before using the techniques and equipment described in this book, it is your responsibility to receive hands-on training under the guidance of a qualified instructor who is experienced in the field of high angle techniques. Such instructors are in established training organizations such as in government organizations and fire training centers or independent schools with a long history of training in high angle techniques. You should choose an instructor or training school with experience in the environment in which you will be working, and with knowledge of the relevant standards and regulations. For example, if you expect to be working in an area that could involve confined spaces, the instructor must understand the government regulations that apply to confined space entry and confined space hazards. An instructor in fire service rescue must be knowledgeable with the relevant NFPA standards.

Instruction in high angle techniques is only the first step. High angle skills deteriorate quickly without constant and regular training. A training schedule should be an indispensable part of every high angle technician's routine. When there is a sudden and unexpected emergency, people react with actions that are instinctive. The only way to make high angle skills instinctive is *practice.*

Common Component Strengths of Typical Technical Rescue Gear*
Approximate Minimum Breaking Strength (MBS)

1-inch (25 mm) tubular webbing pulled end to end	approx. 17.9 kN (4,000 lbf)
1-inch (25 mm) tubular webbing tied in a simple loop	approx. 26.9 kN (6,000 lbf)
1-inch (25 mm) flat webbing pulled end to end	approx. 26.9 kN (6,000 lbf)
NFPA type L carabiner, major axis	> 27 kN (6,069 lbf)
NFPA type L carabiner, minor axis	> 7 kN (1,574 lbf)
NFPA type L carabiner, gate open major axis	> 7 kN (1,574 lbf)
NFPA type G carabiner, major axis	> 40 kN (8,992 lbf)
NFPA type G carabiner, minor axis	> 11 kN (2,473 lbf)
NFPA type G carabiner, gate open, major axis	> 11 kN (2,473 lbf)
NFPA type L ascenders/rope grabs	> 5 kN (1,124 lbf)
NFPA type G ascenders/rope grabs	> 11 kN (2,473 lbf)
Typical $7/_{16}$-inch (11 mm) static rescue rope	> 27 kN (6,070 lbf)
Typical $1/_2$-inch (12.5 mm) static rescue rope	> 40 kN (8,992 lbf)
NFPA light-use static rope	> 20 kN (4,496 lbf)
NFPA general-use static rope	> 40 kN (8,992 lbf)
Strength of the above ropes with the following knots:	
Bowline	70% to 75%
Figure 8 on a bight	75% to 80%
NFPA type L pulley	> 22 kN (4,946 lbf)
NFPA type G pulley	> 36 kN (8,093 lbf)
NFPA type L anchor straps in straight line load	> 22 kN (4,946 lbf)
NFPA type G anchor straps in straight line load	> 36 kN (8,093 lbf)

*MBS provided is for typical rescue gear available from reputable technical rescue equipment suppliers. Components with similar descriptions may have different values. *NFPA*, National Fire Protection Association.

1

Rope Care

INSPECTING A ROPE

After each use, inspect your rope thoroughly by *looking and feeling* along every inch of its length.

Visually inspect the rope, looking for the following:

- *Discoloration:* An obvious change from the rope's original color. Discoloration, particularly brown, gray, black, or green, could indicate chemical damage.
- *Glossy marks:* Could indicate heat fusion damage.
- *Exposed core fibers* (white in most static rope): Indicates damage to the sheath.
- *Lack of uniformity in diameter and size:* May indicate broken sheath bundles.
- *Inconsistency in texture and stiffness* (hold the rope in a loop and see if it is a uniform radius around the entire bend): An inconsistency in the bend may be the result of a soft spot that indicates core damage (Figure 1-1).

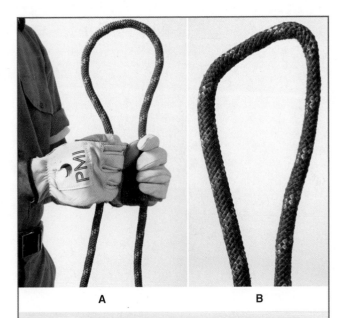

| A | B |

Figure 1-1 Normal and damaged rope being inspected. **A,** Hold the rope in a loop and see if it is a uniform radius around the entire bend. **B,** An inconsistency in the bend may signify a soft spot indicating core damage.

Run the rope slowly through your bare hands, feeling for the following:

- Stiffened fibers
- Obvious changes in diameter
- Soft or hollow spots
- Contamination with dirt and grit

If enough strands are broken, there will be a localized change in the diameter of the rope usually indicated by a depression or hourglass shape that can be felt. Some types of damage will result in "puffs," or core fibers protruding from the sheath. If rope is contaminated with dirt, it should be washed.

GUIDELINES FOR RETIRING A ROPE

The following are general guidelines that can assist in deciding when to retire a rope:

- *Sheath wear:* More than half of the outer sheath yarns are broken.
- *Shock loading:* The rope has been subjected to severe shock loading.
- *Overloading:* The rope has been subjected to overloading for which it was not designed. For a life support rope, this would include towing a vehicle or hauling heavy equipment or materials.
- *Chemical contamination:* Unless the chemical is specifically known to be harmless, consider it a contaminant.
- *Lack of uniformity in texture:* The rope has soft, mushy places or hard spots.
- *Age:* The rope is simply worn out from use.
- *Lack of uniform diameter:* The rope necks down to a smaller diameter in places, similar to an hourglass shape.
- *Loss of faith:* You suspect that the last person to use the rope may not have taken proper care of it.
- The bottom line is always: *When in doubt, throw it out.*

2

Knots and Anchoring

KNOTS

All knots should be dressed (the strands aligned and uncrossed) and compacted (all ends pulled down so that the knot is compact). This ensures that the knot has its greatest holding power, yet as much rope strength as possible is maintained.

Bowline on a Coil

Figure 2-1 shows the tying of a bowline on a coil.

For use with:

- Rope

Purpose:

- To create an improvised litter spider.

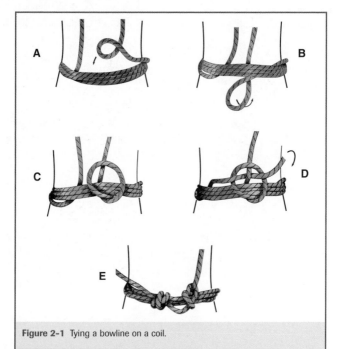

Figure 2-1 Tying a bowline on a coil.

Bowline, Interlocking Long Tail

The interlocking long tail bowline can be used in litter lowering and raising operations. Figure 2-2 shows the tying of an interlocking long-tail bowline.

    For use with:
- Rope

    Purpose:
- To bring the main line and the belay line to a single point, such as the main attachment point on a litter bridle, while providing tails to back up the litter and the litter attendant.

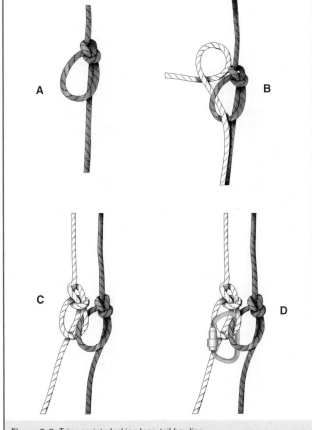

Figure 2-2 Tying an interlocking long-tail bowline.

Clove Hitch

Figure 2-3 shows the tying of a clove hitch. This method uses the working end of the rope, and the hitch can be wrapped around an object that is closed on both ends, such as a litter rail.

For use with:

- Rope
- Webbing

Purpose:

- To anchor to rounded anchor points, such as litter rails.
- To anchor to an object that is closed on both ends.

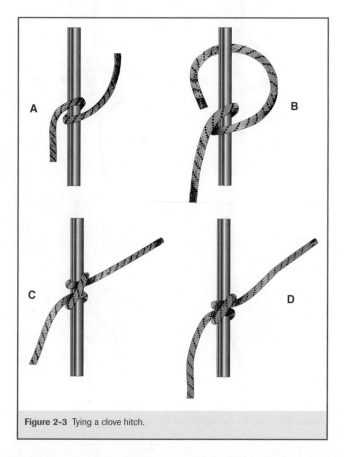

Figure 2-3 Tying a clove hitch.

Double Overhand

The double overhand is essentially one half of a double fisherman's knot.
Figure 2-4 shows the tying of a double overhand knot.

For use with:

- Rope

Purpose:

- To back up other knots.

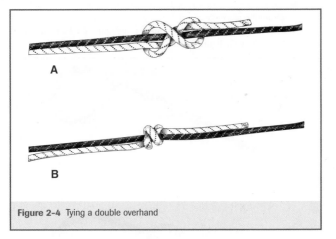

Figure 2-4 Tying a double overhand

Figure 8

Figure 2-5 shows the tying of a simple figure 8 knot.

For use with:

- Rope

Purpose:

- To be used as a stopper knot for certain types of security, such as:
  - Tied in the bottom end of a rope to prevent a person from rappelling off the end.
  - Tied in the top of a rope to prevent it from accidentally slipping through equipment.
  - Used to create a foundation knot for beginning the figure 8 follow-through or the figure 8 bend.

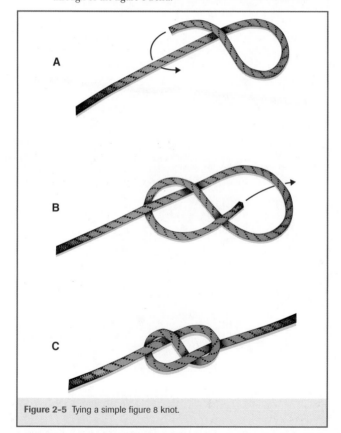

Figure 2-5 Tying a simple figure 8 knot.

Figure 8   11

▨ *Caution*

Do not confuse the figure 8 knot with the simple overhand knot. Compare it with the overhand knot, shown in Figure 2-6. Note the extra step in tying the simple figure 8 knot. When you hold a simple figure 8 knot up by either end of the rope, it should have the rough appearance of the numeral 8.

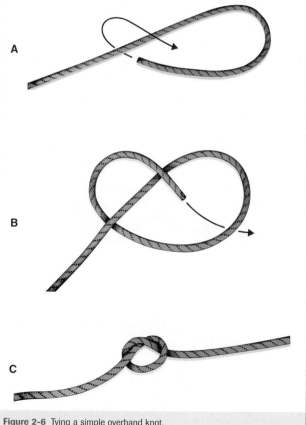

A

B

C

Figure 2-6 Tying a simple overhand knot.

Figure 8 Bend

The term *bend* as applied to knots refers to the joining of two ropes (Figure 2-7).

For use with:
- Rope

Purpose:
- To connect two pieces of rope.
- To create a loop of rope by joining the two ends of one rope.

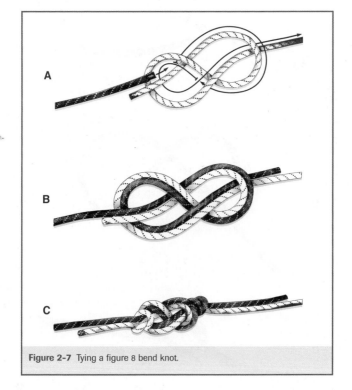

Figure 2-7 Tying a figure 8 bend knot.

Figure 8 Follow-Through Knot   **13**

▨ *Caution*

Note that the figure 8 bend always begins with the step of tying a simple figure 8 knot as a foundation. The next step is to follow exactly the contour of the first knot with the rope ends approaching from *opposite* directions.

Figure 8 Follow-Through Knot

For use in situations where you would want to anchor to an object closed at both ends (such as a structural beam) or to a tall object (such as a tree) when you cannot get a simple loop *over the end* of the object. Instead, you can tie a figure 8 follow-through *around it* (Figure 2-8).

For use with:
- Rope

Purpose:
- To create a loop at the end of a rope in situations where a figure 8 on a bight cannot be tied.

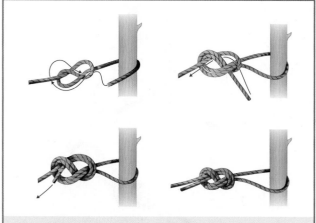

Figure 2-8 Tying a figure 8 follow-through knot.

Figure 8 on a Bight

Note that a *bight* is simply the loop formed when a rope is doubled back on itself. Creating a bight is the first step in tying this knot. You can do this in the middle of the rope or at either end, as shown in Figure 2-9.

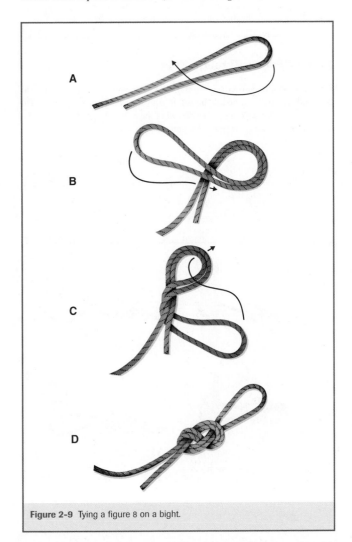

Figure 2-9 Tying a figure 8 on a bight.

Figure 8 on a Bight **15**

For use with:

■ Rope

Purpose:

■ To create a secure loop in a rope for clipping into things, such as:

■ Safety lines

■ Persons being lowered

■ Litter and other rescue equipment

■ Anchor lines

Caution

Be careful to tie a figure 8 on a bight and not an overhand on a bight, which is shown in Figure 2-10.

Figure 2-10 Overhand on bight.

Grapevine Knot (Double Fisherman's Knot)

The grapevine knot, also known as the *double fisherman's knot,* should be used only to join ropes of a similar diameter (Figure 2-11). It should not be used for webbing, for ropes of greatly unequal diameters, or for materials that may tend to untie or creep back through the bends of the knot.

For use with:

■ Ropes of similar diameters

Purpose:

■ To connect two pieces of rope.

■ To create a loop of rope by joining the two ends of one rope.

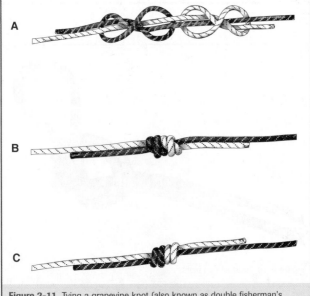

Figure 2-11 Tying a grapevine knot (also known as double fisherman's knot).

Münter Hitch

The Münter hitch often is used in the belaying of one-person loads. It sometimes is used as an emergency rappel device, when a rappeller has no other descent device. It can be used as an element of load-releasing hitches, such as the radium-releasing hitch. Although there are several ways of tying a Münter hitch, Figure 2-12 shows a simple, easy to remember method.

For use with:

- Rope

Purpose:

- To belay one person loads.
- To be used as an emergency rappel device when no other descent device is available.
- To be used as an element of load-releasing hitches.

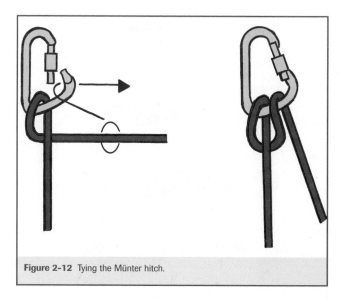

Figure 2-12 Tying the Münter hitch.

Ring Bend (Water Knot, Overhand Bend)

The ring bend knot, also known as the *overhand bend* or *water knot,* is used for webbing only. Figure 2-13 shows the tying of a ring bend in webbing.

For use with:

- Webbing

Purpose:

- To join two different pieces of webbing to form a longer piece.
- To tie the two ends of one piece of webbing together to form a loop.

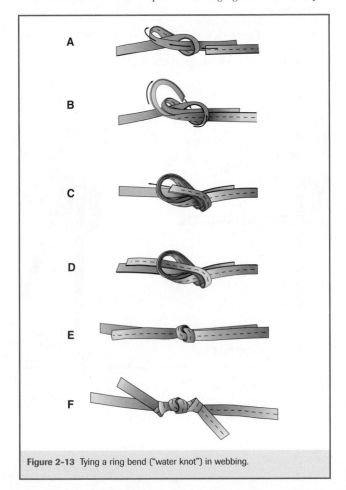

Figure 2-13 Tying a ring bend ("water knot") in webbing.

▨ *Warning*

The ring bend is to be used only for webbing. Do not use it for rope. Because of its flat nature, webbing tends to contour over itself. Rope does not have this quality, and a water knot in rope may easily come out.

▨ *Warning*

Always have at least 2 inches (5 cm) of webbing in the ends of ring bends after they have been tied and pulled tight. Although webbing contours well in a ring bend, it tends to be slippery. For additional insurance, back up both sides of the knot with an overhand knot (see Figure 2-13, *F*, p. 18) or sew loose ends down sufficiently to keep them from working through the knot.

A ring bend in webbing should be inspected frequently, because over time it tends to work loose.

Make sure the webbing follows flat through the knot. A twist in the webbing inside the knot will allow the knot to slip at relatively low loads.

ANCHORING TECHNIQUES

Tensionless Hitch (High-Strength Tie-off)

The total number of wraps depends on the diameter of the anchor point and the coefficient of friction of the anchor point surface (Figure 2-14).

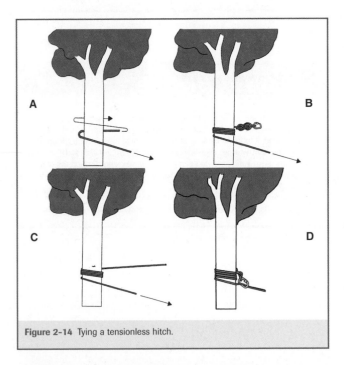

Figure 2-14 Tying a tensionless hitch.

The tensionless hitch works only if there is adequate surface contact between the rope and the object. If the shape of the object does not allow this, choose another method of anchoring to it. For example, an H-shaped beam has rope contact in only four places; a tensionless hitch will not work in this case.

If the anchor point has sharp edges, as are sometimes found on structural beams, pad the rope appropriately.

Wrap 3, Pull 2 for Anchoring

In this system, there are three wraps around the anchor with two loops pulled out. Note that for maximum strength the webbing knot is on the interior loop, where there will be the least stress on it (Figure 2-15).

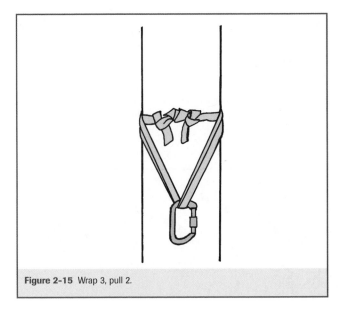

Figure 2-15 Wrap 3, pull 2.

Establishing a Picket System

1. The pickets should have a minimum length of 5 feet (1.5 m) so that there will be a minimum of 3 feet (1 m) in the ground and a maximum of 2 feet (0.6 m) above ground (Figure 2-16).
2. Drive the pickets at an angle of 15 degrees away from the force to be anchored.
3. Connect the pickets in each row by lashing from the top of the first picket (the one closest to the load) to the bottom of the next picket, close to the ground. Continue in this manner until all the pickets in the row have been lashed together.
4. Tension the lashings by twisting with a stick four to six turns. Drive this stick into the ground to secure it.
5. Construct the next rows of pickets as described above.
6. Connect the main line by clipping it to the front picket in each row with a load-sharing anchor system as described above.

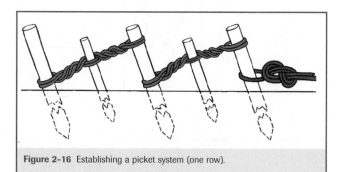

Figure 2-16 Establishing a picket system (one row).

Pretensioned Back-Ties

A pretensioned back-tie is a method of removing slack in a back-tied anchor system before it is loaded (Figure 2-17).

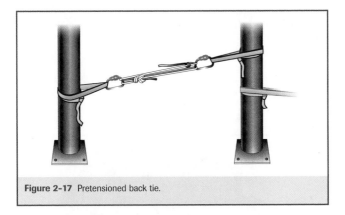

Figure 2-17 Pretensioned back tie.

One technique for creating a pretensioned back-tie is to use rope to connect two anchor points that are tensioned with a simple mechanical advantage (MA) system. To construct the system in Figure 2-17:

1. Place wrap three, pull two webbing attachments on each anchor facing one another.
2. Place a large locking carabiner in each of the web attachments.
3. Use a length of rope to create a simple 3:1 MA system using the carabiners.
4. Start by attaching the rope to the carabiner at the rear anchor.
5. Run the rope through the forward carabiner, back through the rear carabiner, and pull it forward until it almost reaches the front carabiner.
6. Tension the simple haul system using two people. After the initial tensioning, "vector" the system by pulling it back and forth sideways to remove additional slack.
7. While maintaining the tension, tie off the system near the front carabiner using half hitches.

NOTE: The pretensioned back-ties should be close to in line with the direction of pull of the lowering or haul system or highline. They should never be more than 15 degrees off to either side.

Load-Sharing Anchors

The simplest way to create a multiple anchor system for load sharing is to use two anchor ropes or slings of equal length. Run them from different anchor points and clip them together into a single point using one or two large locking carabiners (Figure 2-18). The point where you clip the two lines together with the carabiners is known as a *master attachment point*. In load-sharing anchors, the master attachment point is where the multiple lines come together.

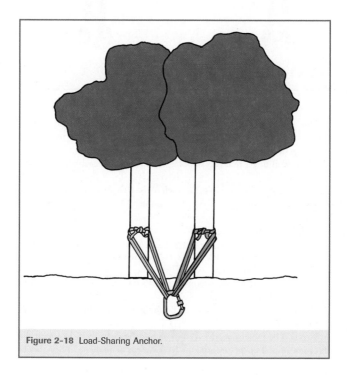

Figure 2-18 Load-Sharing Anchor.

Warning

A primary concern in rigging a load-sharing anchor is not to create too wide an angle between the legs of the anchor system. Ideally, this angle should not exceed 90 degrees, and it must *never* exceed 120 degrees. Beyond this point, the forces on each anchor are greater than the total load itself.

Remember, any angle in an anchor system increases the loading on anchors and other elements of the system. Only when the angle between the legs of the anchor system is 0 degrees does each leg carry half the load. The diagrams in Figure 2-19 show how angles affect the forces on anchor points and other elements of the system.

After you finish rigging the anchor system, check all the carabiners to make sure they will load correctly.

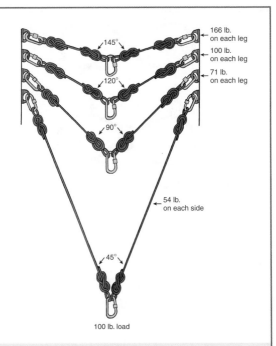

Figure 2-19 Relationship between anchor sling tension and a 100-lb load at different angles. 145°; 166 lb on each leg. 120°; 100 lb on each leg. 90°; 71 lb on each leg. 45°; 54 lb on each leg, 100 lb load.

Simple Two-Point, Load-Distributing Anchor (Figure 2-20)

Configure a loop of webbing or rope in the shape of an 8.

Clip a large locking carabiner across the inside loop.

Take each end of an outside loop and clip it into an anchor point.

Clip the carabiner on the inside loop into the main line; this will be the focal point for the anchor system.

Make sure the angles made by the sling do not exceed the critical ones described in the Warning box on p. 25.

Before loading this system, test it by hand. One at a time, simulate failure of each of the anchors to make sure the system will catch the load.

Whatever the direction of the pull, the central carabiner should slip along the sling to equalize the forces. Should one anchor point fail, the webbing should set itself to pull on the other anchor point.

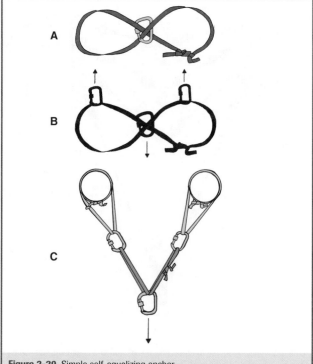

Figure 2-20 Simple self-equalizing anchor.

Two-Loop Figure 8, Load-Distributing Anchor System (Figure 2-21)

Create a large loop; use a length of rope and tie the two ends together with a figure 8 bend (or double fisherman's knot).

For ease of operation, set the circle of rope on the ground or the floor.

Place the knot at about 3 o'clock or 9 o'clock on the circle so that it stays out of the way.

Take a large bight of rope and flip it back inside the circle, about two thirds of the way up.

The lower section of the circle is now doubled, so that there are four strands of rope.

Gather together the four strands and tie a figure 8 knot with them.

At the top of the circle, there is now a large loop with a smaller loop inside.

To reduce shock loading, keep the size of the large loop to a minimum,

Figure 2-21 Complex self-equalizing anchor.

with a circumference of no more than 8 feet (2.4 m). At the bottom of the circle is a much smaller loop.

Take the larger loop at the top of the circle and clip it into all the anchor points using a locking carabiner at each point.

Take the smaller loop at the bottom of the large loop. Using locking carabiners, clip this together to the larger loop between each anchor point.

Take the small loop below the knot created from the four strands. Clip this into the main line using one or two large locking carabiners. This will be the focal point for the anchor system.

If tied correctly, should any one point fail, the system should redistribute the load among the remaining anchor points. The system should also distribute the load among all points, regardless of the direction from which the load is coming.

Before the system is loaded, the riggers should inspect it to make sure it will perform in this manner. If any problems are noted, such as a knot jamming, the system must be adjusted so that it will work as intended.

▨ *Warning*

Load-distributing anchor systems have the potential for dangerous shock loading when one anchor point pulls out and others take the load. Reduce potential shock loading by keeping angles small and slack to a minimum. Also:

❑ Try to keep anchor points close to one another. If this is not possible, it may be better to extend far away anchor points with static rope to keep the load-sharing anchor's loop as small as possible.

❑ Keep the outside angle to less than 90 degrees; even better, keep it to 60 degrees. This limits the forces on the remaining anchor points if one of the inside anchors fails in a load-distributing anchor with three or more points. The outside angle is measured from the two outermost anchor points down to the master attachment point.

❑ Rig load-distributing anchors with a minimum of slack in the system. Keep rope or webbing length to a minimum, approximately 8 feet (2.4 m) or less, in the tied loop of a three-point system. Rig for a maximum 1-foot (0.3 m) drop with failure of an anchor point.

3

One-Person Techniques

Belaying

Belaying Signals

When you are belaying, it is essential that you use the standard voice signals (also called commands or calls). They take place as an exchange of signals between a climber (or rappeller) and the belayer to ensure that both are ready.

The standard belay signals, in sequence, are as follows:

Signal	Phrase Meaning
Climber: "On belay?"	"I am about to climb (or rappel), are you ready to catch me if I fall?"
Belayer: "Belay on."	"I am ready to catch you if you fall."
Climber: "Climbing (or "Rappelling").	"I am starting to climb (or rappel)."
Belayer: "Climb on (or "Rappel on").	"Go ahead."

Once a climber is in a place where he or she no longer needs the belay, the climber initiates an exchange to end the belay.

Signal	Phrase Meaning
Climber: "Off belay."	"I am in a secure place now. I no longer need the belay."
Belayer: "Belay off."	"I am no longer belaying you."

When the climber and belayer are not within sight of one another, they can extend this sequence to include an "Off rope" statement by the climber (which is repeated by the belayer). This helps the belayer to know when to start pulling rope, when he or she should hang onto the rope because it might disappear, or when the rope has been cleared for other needs.

Signal	Phrase Meaning
Climber: "Off rope!"	"I am finished with the rope."
Belayer: "Off rope."	"I understand that you are finished with the rope."

Some additional signals can help communication between climber and belayer during the belay cycle. One of these is in response to a situation in which the belayer is holding the rope too tightly.

Signal	Phrase Meaning
Climber: "Slack"	"There is too much tension on the rope; I cannot move as well as I would like."
This signal requires no verbal response from the belayer, only the action of letting an appropriate amount of slack into the rope.	

Another situation involves the opposite problem: too much slack in the rope. It may be that the climber is at a particularly tricky point and he or she needs the support of the rope to make a move.

Signal	Phrase Meaning
Climber: "Tension."	"Hold the rope tightly for a bit; this might be a difficult move."
This signal requires no verbal response from the belayer, only the action of taking slack out of the rope.	

Tying the Münter Hitch for Use in Belaying

See Knots, p. 17.

Rappelling

Warning

The learning and practice of rappelling techniques must be done under the guidance of a qualified instructor. Rappelling techniques must be practiced first on level ground and then on short and moderate slopes before being used on a steep face. Everyone practicing rappel techniques on a steep face or any other area where a severe fall is possible must use a top belay.

Rappelling with the Figure 8 with Ears

Figure 3-1 shows the procedure for lacing up a figure 8 with ears for rappelling.

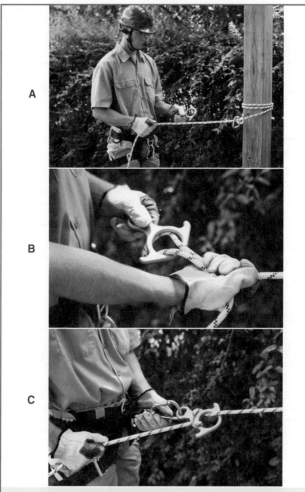

Figure 3-1 Using figure 8 on level ground. *Continued*

Figure 3-1, cont'd

Locking Off the Figure 8 with Ears

Figure 3-2 shows the procedure for locking off a figure 8 with ears.

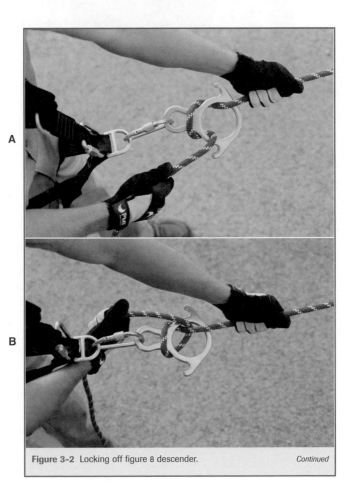

Figure 3-2 Locking off figure 8 descender. *Continued*

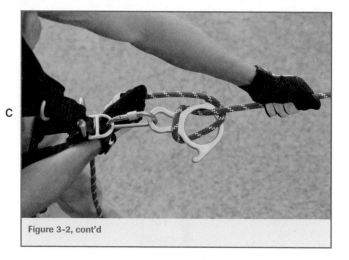

C

Figure 3-2, cont'd

Gaining Extra Control Using a Figure 8 with Ears

Double-Wrapping

Figure 3-3 shows the procedure for double wrapping a figure 8 with ears to gain extra friction.

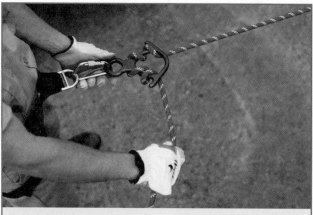

Figure 3-3 Hand position with double wrapped 8.

Using a Spare Carabiner

Figure 3-4 shows the sequence for using a spare carabiner with a figure 8 with ears to gain extra friction.

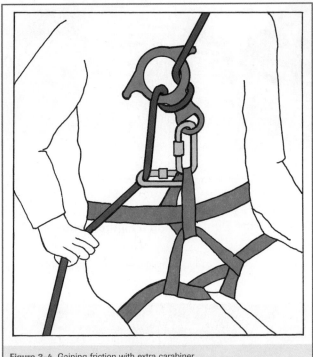

Figure 3-4 Gaining friction with extra carabiner.

Rappelling with the Brake Bar Rack

Figure 3-5 shows the procedure for attaching a brake bar rack to a harness and rope.

Figure 3-5 Brake bar rack on harness.

Ascending

Selecting Material for a Prusik Hitch

The Prusik hitch operates more efficiently if it is made from a rope that is smaller in diameter than the main-line rope to which it is attached. As a rule, the diameter of the Prusik cord should be two thirds to three fourths that of the main-line rope. This usually works out to using 8-mm Prusik cord on $^{7}/_{16}$-inch (11 mm) rope and 9 mm Prusik cord for $^{1}/_{2}$-inch (12.5 mm) rope. Avoid extremes. The Prusik cord must be strong enough to support the intended load with a proper safety margin. However, it should not be so large that it is hard to get the hitch to "set."

Creating a Prusik Loop

To create a Prusik loop for a seat harness, make a continuous loop from a length of cord, approximately 6 feet (2 m), that you have chosen for the loop material. Do this by tying the two ends together with a grapevine (double fisherman's) knot.

▨ *Warning*

Materials used for Prusik hitches wear quickly and should be inspected before each use.

Attaching a Prusik Loop to a Rope

Figure 3-6 shows how to tie a two-wrap Prusik onto a rope. This illustration shows the rope anchored vertically, but a Prusik can be tied onto a rope at any angle.

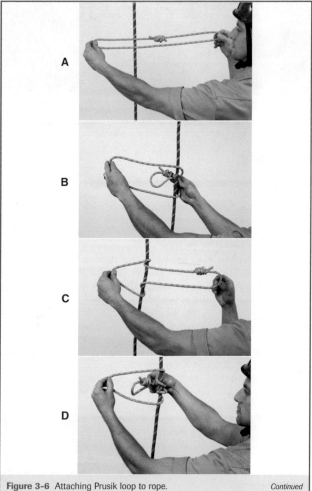

Figure 3-6 Attaching Prusik loop to rope.

Continued

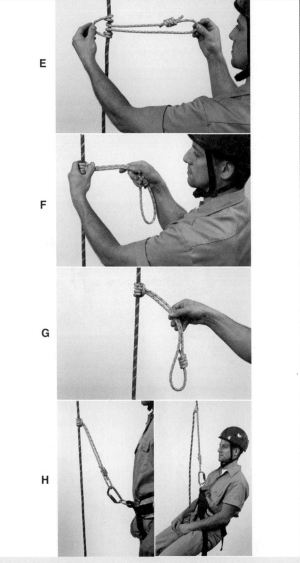

Figure 3-6, cont'd

Tying Off Short

Tying off short (Figure 3-7) is the safety procedure of tying directly into the main rope to ensure an additional attachment. It is used in certain situations during ascending when there is a danger of the person losing attachment to the rope, such as when:

- A person is using only two ascenders and must take one ascender from the rope for a procedure such as moving past a knot or going over the edge of a cliff or building.
- A knot must be crossed on the way up the rope.
- Going over an edge or obstruction and the climber does not have a quick attachment point. (A quick attachment point, also known as a *quick attachment safety,* is a short sling attached on one end to the

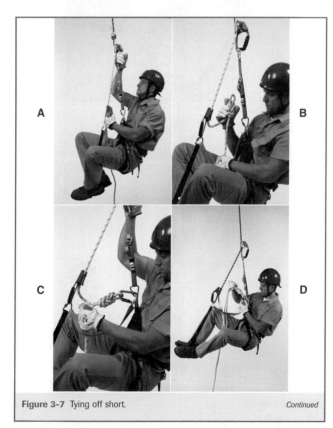

Figure 3-7 Tying off short. *Continued*

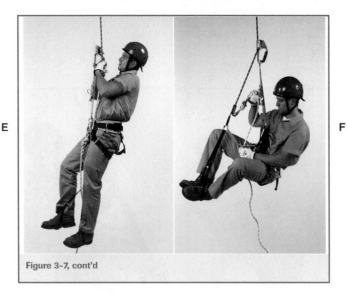

Figure 3-7, cont'd

user's seat harness and on the other end to an ascender, which can be easily attached with one hand to a secure point.)

- Using an ascending system that cannot guarantee the user will remain upright if one ascender fails. (A well-designed ascending system has at least two points of contact above the climber's center of gravity so that if either ascender fails or comes off the rope, the climber ends up sitting in his or her harness and not hanging upside down by a heel).

Changing Over

Changing over means switching from ascending to rappelling, or from rappelling to ascending, while still on the rope. It is a skill that adds to vertical competency and is particularly useful in emergency situations, such as when a person rappels to the end of a rope and finds that it does not reach the bottom.

Equipment Needed for Changing Over

- Two locking carabiners in the seat harness tie-in point (the one in use at the time and a spare to be used when changing over).
- Harness gear loops or a gear sling. The equipment not in use at the time (i.e., the rappel device while ascending, the ascenders while rappelling) is attached to these.
- Equipment for both ascending and rappelling.

Changing Over from Ascending to Rappelling

Figure 3-8 shows example procedures for changing over from ascending to rappelling.

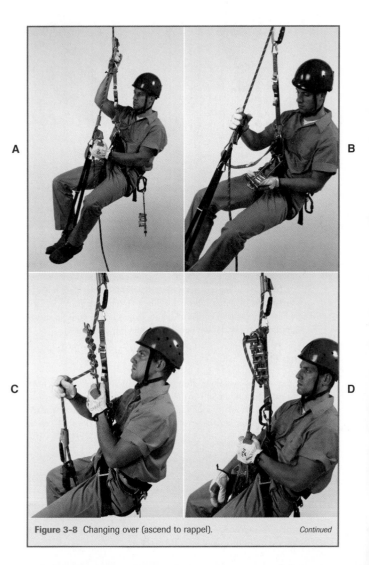

Figure 3-8 Changing over (ascend to rappel). *Continued*

E

F

G

H

Figure 3-8, cont'd

Changing Over from Rappelling to Ascending

Figure 3-9 shows example procedures for changing over from rappelling to ascending.

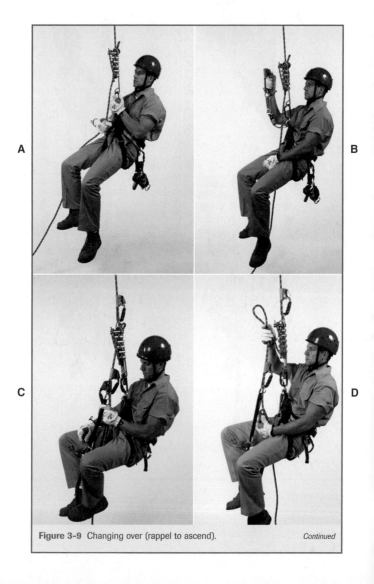

Figure 3-9 Changing over (rappel to ascend). *Continued*

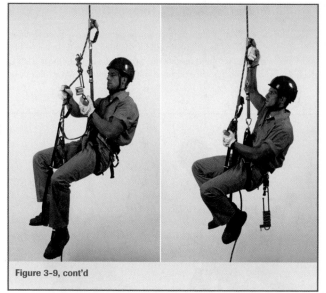

Figure 3-9, cont'd

Extricating from a Jammed Rappel Device

The ability to extricate from a rappel device jammed by hair or clothing without using a knife is an essential skill for the high angle technician. The skills and equipment required for this procedure are similar to those used in changing over. Because of the real possibility of a jammed rappel device or

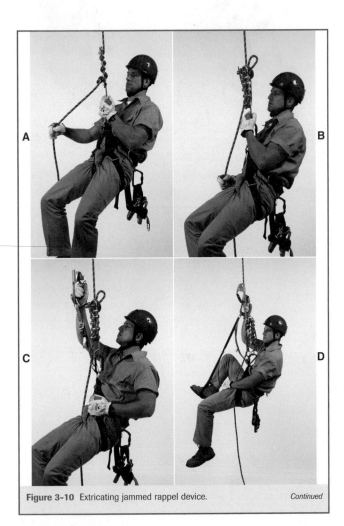

Figure 3-10 Extricating jammed rappel device.   *Continued*

similar emergency occurring, it is wise to carry the following spare equipment when rappelling:

- Two ascenders, any type with connecting slings, or two Prusik loops. The length of the Prusik loops varies according to body proportion.
- A spare, large-locking carabiner.

Figure 3-10 shows an example procedure for using ascenders to extricate from a jammed rappel device.

Figure 3-10, cont'd   *Continued*

Figure 3-10, cont'd

Caution

If you are using a brake bar rack, make sure that when you lock it off, the brake hand rope does not get trapped below the top bar (Figure 3-11). It must remain above the top bar, or the device will jam when the rope tension comes onto it.

Figure 3-11 Avoid trapping rope below top bar.

4

Communications Coordination

Whistle Blasts for Communication

A nonelectronic alternative for communication is the use of whistle blasts. Obviously these are limited in the degree of information that can be conveyed, but they often are audible where nothing else works. The exact form of these whistle communications must be worked out and used in practice in advance. *SUDOT* is a recognized whistle command system used in rope rescue.

S—Stop (one blast)
U—Up (two blasts)
D—Down (three blasts)
O—Off rope (four blasts)
T—TROUBLE! (long, continuous blast)

Organization and Management

Key Points in Maintaining an Organized Rescue

- **Initiate a quick "size up" of the incident.** To verify the initial report and scope of the incident, send the closest available resource directly to the scene.
- **Organize an immediate initial response to reach and stabilize the rescue subject.** Stabilize the subject medically by treating life-threatening injuries. Physically stabilize the individual so that they cannot fall farther. Psychologically stabilize the subject with professional reassurance.
- **Use the incident command system/incident management system (ICS/IMS).** Identify the command positions of rescue personnel, verbally on the radio and through the use of vests.
- **Establish an accessible staging area for equipment.** This area must not impede the area used for operational rigging.
- **Limit communications with the rescuers in technical terrain to the edge manager or the operations chief.** Working in the vertical realm is complicated. Rescuers' efficiency is compromised when they are overtasked with too many instructions.
- **Stay ahead of the logistics curve.** Plan and act now. Be prepared for a rescue to take longer than expected. Request additional resources, supplies, and equipment well in advance so that efficiency does not suffer.
- **Keep rescue systems simple and safe.** An overly complex system may compromise efficiency.

Important Safety Reminders

- Do not rush! Maintain a sense of *controlled urgency*.
- Use well-trained, competent rescuers for the core of the team. Use the most highly skilled rescuers in decision-making roles and where technical

competence is absolutely essential. Less experienced or less-skilled personnel can easily fill roles as low angle litter bearers, hauling team personnel, or equipment runners.

- Rescuers should be prepared for contingencies. Have adequate personal gear for staying warm, dry, fed, and well hydrated. *Aggressively* use appropriate personal protective equipment (PPE) for *all* incident hazards, environments, and tasks (e.g., gloves, footwear, helmet, harness, personal flotation device, hearing protection, Nomex clothing, safety goggles, sunscreen). Have spare equipment available. Wear helmets in the rock fall zone.
- Establish a well-marked safety perimeter. Consider flagging or a chalk line, as well as the use of chemical light sticks at night to identify this line.
- Stay tied-in when working within 10 feet (3.3 m) of an exposed edge. Establish a marked safety perimeter with flagging, chalk, or chemical light sticks.
- Minimize the number of personnel working near an exposed edge.
- Designate a incident safety officer (ISO), which might be a collateral role for a rescuer.
- Do a safety check before using a system. Recheck equipment when in use.
- Engineer a *redundant* system. Rescue systems have backups.
- Use standard communications terminology and techniques.
- Use edge protection to protect lines from sharp rock edges.
- Establish safety lines where exposure places personnel at significant risk of injury.
- Make a point of not standing inside a rope bight under tension during a raising operation. Prevent personal injury in the event a change of direction fails in a hauling system. Stand to the outside of the rope bight.
- Never get on a rope without adequate gear to go up the rope and down the rope. If you are rappelling, make sure you have ascenders with you.
- Secure loose gear in a cache adjacent to the rescue operations area.
- Keep a Prusik and trauma scissors with you to handle a self-rescue situation.
- Remain alert to prevent cross-gate forces and three-way forces on carabiners.
- Derig gear after the rescue, starting at the cliff edge and working away from it.

5
Litters

Litters and Rescue Subject Packaging

Improvised Litter Tie-In for a Rescue Subject

Rescuers commonly use 1-inch (25 mm) tubular webbing to secure a rescue subject in the litter. The following litter tie-in system has two components: an upper torso component and a lower torso component. The upper torso component uses two pieces of tubular webbing: each 1 inch (25 mm) wide and 30 feet (9 m) long. The lower torso component uses one piece of tubular webbing: 1 inch wide and 30 feet (9 m) long (Figure 5-1).

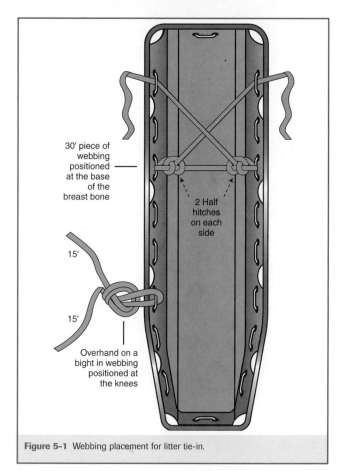

Figure 5-1 Webbing placement for litter tie-in.

Anchoring Rescue Subject Tie-Ins to the Litter

Avoid running tie-in webbing over the top rail of the litter, where it is subject to abrasion and cutting, which could weaken the tie-in. Also, tie-in webbing on the top rail can interfere with other rescue rigging, such as bridles or spiders, if these also will be attached there.

Upper Torso Component (Figure 5-2)

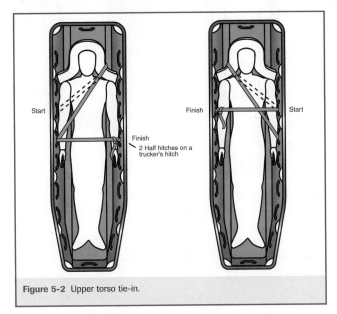

Figure 5-2 Upper torso tie-in.

Lower Torso Component (Figure 5-3)

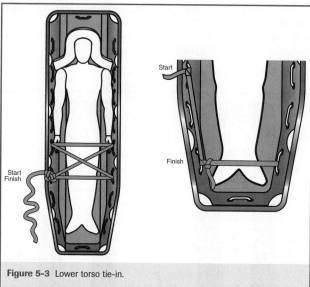

Figure 5-3 Lower torso tie-in.

⫸ *Warning*

- ❑ Do not lash webbing horizontally across the upper chest or neck. Should the subject slide down in the litter, a line of webbing across this area could strangle the person.
- ❑ If the webbing is too tight, prolonged loss of circulation could result in serious medical problems, such as compartment syndrome. In extreme cold, reduced circulation can increase the potential for frostbite or burns from rewarming sources such as hot water bottles or heating pads. Pad pressure points created by tie-ins. Check the subject's circulation after completing the tie-in. Recheck circulation regularly.
- ❑ Litter tie-ins can work loose; constantly monitor the litter lashing.

Tying the Main-Line Rope Directly to the Head of the Litter

If the low angle evacuation is only one rope length and the rope will not have to be detached from the litter during the operation, the main-line rope may be tied directly onto the head of the litter. Figure 5-4 shows a typical system for attaching a main-line rope to the head of a litter for low angle evacuation. This attachment consists of a very large loop created at the end of the rope by an end knot, such as the figure 8 follow-through knot.

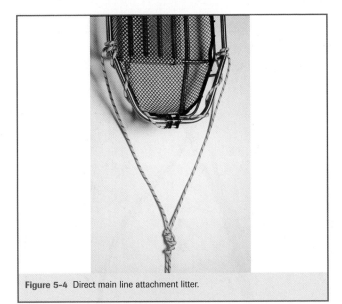

Figure 5-4 Direct main line attachment litter.

Warning

When a main-line lowering or hauling rope is connected to the end rail of a litter, the force must be spread out evenly along the rail by weaving the attachment around the rail multiple times. Never attach a main-line lowering or hauling rope to a single point on the rail of a litter. Many litters are susceptible to failure if sudden forces pull at a single point on the rail. This is particularly true of litter rails that are butt welded. A sudden force at such points can break the weld, causing the litter rail to fail.

Tying a Closed Loop Directly to the Head of the Litter

During low angle evacuations of more than one rope length or when the litter must be quickly freed from the system for a trail carry, the lowering rope must be removed and reattached to the litter. In such cases it is more practical to leave a closed loop of rope tied at the end of the litter. To attach the lowering and hauling line to the loop, tie a figure 8-on-a-bight knot in the end of the main-line rope and clip it to the loop with a large locking carabiner (Figure 5-5).

Figure 5-5 Tying loop onto litter.

▨ *Caution*

The angle made in the loop left tied at the end of the litter, when it is attached to the main-line rope, must not be more than 90 degrees. If the angle is more than 90 degrees, make a larger loop.

Adjustable Litter Tender Tie-in

An option to the fixed length attachment is an adjustable length tie-in (Figure 5-6). An adjustable length tie-in allows the attendant to change the distance to adapt to varying circumstances, such as changing terrain and obstructions. It also is an advantage if different team members must use it.

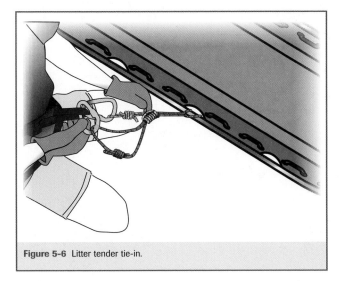

Figure 5-6 Litter tender tie-in.

6

Lowering, Belaying, and Hauling

Low Angle Litter Evacuation

Voice Communications for Low Angle Lowering

Signal	Phrase Meaning
Litter captain to brakeman: *"On rope?"*	The litter team is ready. They should not proceed without a reply from the brakeman.
Brakeman to litter captain: *"On rope!"*	The braking system is engaged and ready to be loaded.
Litter captain to brakeman: *"Down slow"* or *"Down fast."**	The brakeman allows rope through the brakes slowly or faster, per the litter captain's request.
Litter captain to brakeman (this signal may be used by anyone who sees danger or potential problems developing): *"Stop!"*	All rescuers stop activity.
Litter captain to brakeman (or vice versa): *"Stop! Stop! Why Stop?"*	This signal is given when, for an unexpected reason and without command from the litter captain, the rope has stopped moving. It could be that the brakeman is still letting out rope, but the rope is jammed somewhere or the litter is on a ledge. This, and any "Stop" command or unintelligable command, should result in immediate setting of the brakes (or belay, if applicable.)
Brakeman to litter captain: *"Zero!"*	This signal means that only about 20 feet (6 m) of rope is left. The litter team should set the litter down at a convenient spot so that a new brake and anchor set can be established.
Litter captain to brakeman (and belayer when used): *"Off rope."*	The litter has been set down in a secure spot. It and the litter team are in no danger of falling.
Brakeman to litter captain: *"Off rope."*	Acknowledgment of litter captain's signal.

*NOTE: The "Down slow" or "Down fast" signal should be repeated by the brakeman back to the litter captain so that the captain knows the brakeman understands. Otherwise, the litter may be lowered at a different rate from the one desired.

1:1 Mechanical Advantage Hauling System

One of the simplest hauling systems used in low angle evacuation is the *1:1 hauling system*. In essence, the 1:1 ratio simply means that the force needed to haul the load (litter, subject, and attendants) is about the same as the weight of the load. Figure 6-1 shows the elements of a basic 1:1 hauling system.

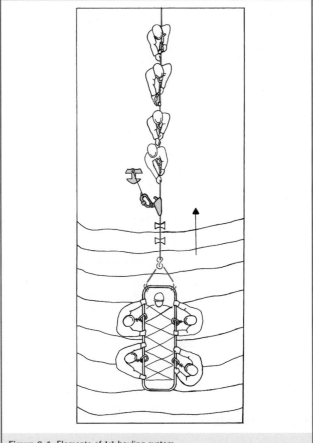

Figure 6-1 Elements of 1:1 hauling system.

▨ *Warning*

Anchors established for pulley directionals with small interior angles must be stronger than if they were simply supporting a weight equal to the load being hauled. The smaller the interior angle ("A") in the rope created at the change in direction, the greater the force on the pulley and its anchor system (Fig. 16-9). Angles less than 120 degrees will place forces on the directional anchor greater than the actual load being hauled.

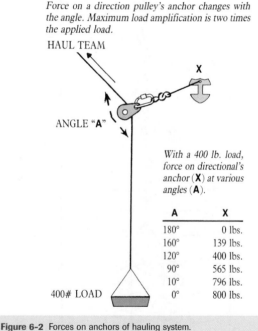

Force on a direction pulley's anchor changes with the angle. Maximum load amplification is two times the applied load.

HAUL TEAM

X

ANGLE "**A**"

*With a 400 lb. load, force on directional's anchor (**X**) at various angles (**A**).*

A	X
180°	0 lbs.
160°	139 lbs.
120°	400 lbs.
90°	565 lbs.
10°	796 lbs.
0°	800 lbs.

400# LOAD

Figure 6-2 Forces on anchors of hauling system.

Haul Commands

To avoid potentially disastrous confusion, haul commands are limited to a few standardized terms.

Signal	Phrase Meaning
	The voice communications for hauling movement are initiated by the litter captain. Other than one special exception, no one else initiates communications for hauling; everyone else remains quiet.
"Haul" or *"Haul slow."*	"Begin hauling." This signal needs to be given only once; the team will continue to haul until given another command is given. "Haul slow" is a variation used when there is an emphasis on slow movement, such as when the litter is about to reach the top.
"Set."	The haul team immediately stops hauling and gently eases back on the load to set the progress capture device. This signal may be given for safety reasons or, more commonly, to get another bite of the rope.
"Slack."	The progress capture device is set, therefore the haul team slacks on the rope. This allows any part of the system to be reset and the haul team to get another bite on the rope. (If a belay is used, the normal belay communications are also used.)
"Stop!"	All movement stops immediately. The haul team holds tension until told to do otherwise. This one command may be given by anyone who sees something going wrong or a problem developing.

Using Progress Capture Devices in Steep Slope Evacuation

Progress capture devices should always be connected to a safe and secure anchor system that is, if possible, separate from the anchor system supporting the haul system. Some models of general-use ascenders have an arrow with the caption "Up." This is the indicator for the direction of use when ascending a rope. *In a hauling system, this arrow should always point along the rope toward the load (the litter and the rescue subject).* Some older model general-use ascenders do not have this arrow inscribed on the shell. However, if you look at the shell in profile, you will see that it vaguely resembles an arrowhead (see Figure 6-3, p. 68). Again, the arrow should point in the direction of the load.

Positioning the Progress Capture Device (PCD)

The progress capture device should be on an anchor that is separate from the hauling system. However, it should be close to, and parallel to, the main-line rope. This helps prevent shock loading and reduces slack, which interferes with haul system efficiency.

Setting the Progress Capture Device

There are two primary concerns when rigging the PCD: (1) the device must grab the rope when needed and (2) the device must not ride up the rope as the rope moves, because this could result in dangerous shock loading. How the PCD is specifically rigged depends partly on the specific type of general-use ascender used and partly on the specific circumstances of the rigging (Figures 6-3 and 6-4).

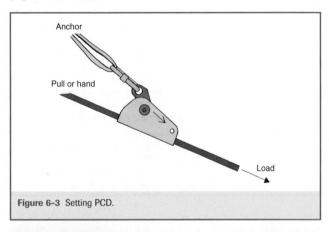

Figure 6-3 Setting PCD.

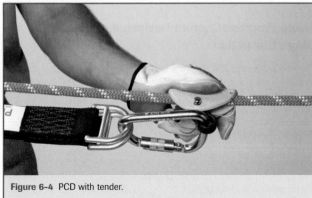

Figure 6-4 PCD with tender.

Free Running Progress Capture Device

Figures 6-4 and 6-5, *A,* show one method of ensuring that the PCD stays in place and clamps on the rope when needed. *This technique is specific to the use of a general-use ascender as a PCD.* The technique requires the services of a person known as the *PCD tender.*

A second technique, using a free running general-use ascender as a PCD, is shown in Figure 6-5, *B.* A bungee (elastic) cord is clipped into the empty hole usually found toward the "point" of the arrowhead in some general-use ascenders. The other end of the bungee cord is anchored securely to a convenient spot toward the load. A great deal of tension is not required on the bungee cord, but there should be enough to (1) keep the PCD from riding up the rope and (2) keep the PCD closed on the rope.

NOTE: Tie the bungee cord *only to the shell* of the general-use ascender, not to the cam itself. Otherwise, the system will not function as needed.

A third alternative for setting a general-use ascender PCD is shown in Figure 6-5, *C.* This method uses a short piece of cord attached to a weight. The weight is hung so that it will pull the shell of the PCD toward the load.

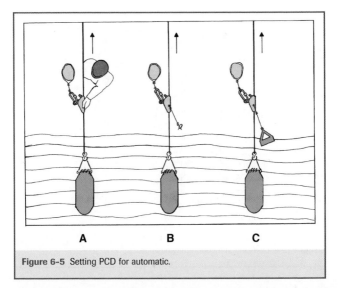

Figure 6-5 Setting PCD for automatic.

Counterbalance Haul System

Strictly speaking, a counterbalance hauling system is also a 1:1 haul system. Without considering loss due to friction and other inefficiencies, the force used to haul is the same as the weight of the load being hauled. Figure 6-6 shows the elements of a counterbalance haul system.

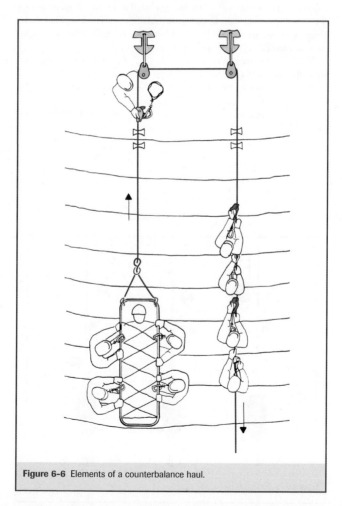

Figure 6-6 Elements of a counterbalance haul.

2:1 Hauling System for Low Angle Evacuation

Figure 6-7 shows a 2:1 hauling system for low angle evacuation.

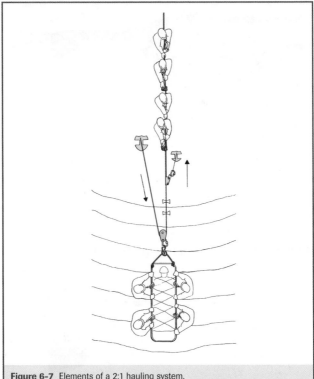

Figure 6-7 Elements of a 2:1 hauling system.

▧ *Warning*

Although a 2:1 haul system can make hauling easier, it does increase rope management problems:

- ❏ Two strands of rope are now moving along the same path.
- ❏ A pulley is moving, which could easily become jammed on underbrush, trees, or rock. Rescue personnel must be able to reach this pulley, wherever it jams, to free it.

Rescue Belaying

Tandem Prusiks (Figure 6-8)

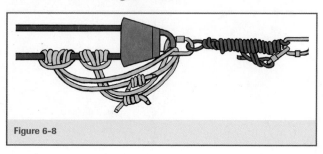

Figure 6-8

The following procedure assumes that a ½-inch (12.5 mm) rope is used for the belay line.

Using approximately 10 feet (3 m) of 8 mm or 9 mm nylon climber's accessory cord, cut two lengths:

- 5.5 feet (1.67 m)
- 4.4 feet (l.34 m)

Radium-Releasing Hitch (Figure 6-9)

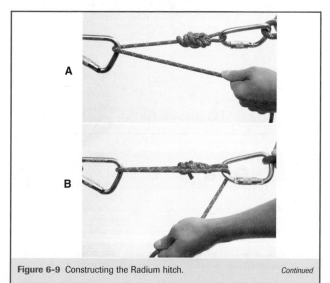

Figure 6-9 Constructing the Radium hitch.   *Continued*

Materials needed

- Two locking carabiners of suitable strength for rescue loads
- 33 feet (10 m) of ⁵⁄₁₆-inch (8 mm) nylon static kernmantle rope or 8 mm nylon accessory cord

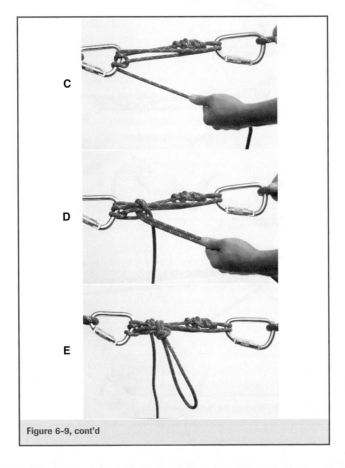

Figure 6-9, cont'd

Releasing the Radium-Releasing Hitch (Figure 6-10)

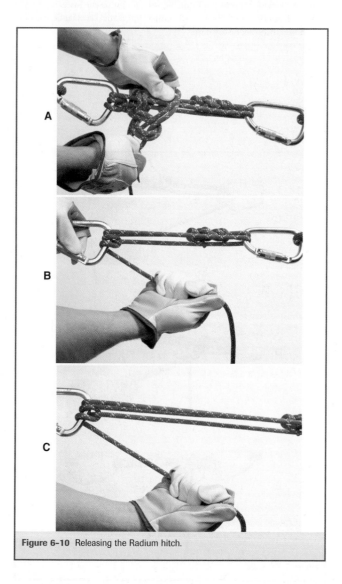

Figure 6-10 Releasing the Radium hitch.

- ❑ Never release any load-releasing hitch until you are certain the load can be successfully transferred to some other system or will reach the ground before all the cord in the load-releasing hitch has been let out.
- ❑ Make sure you will be able to control the load as you release it. Extreme loads may require two or more people.
- ❑ Keep your fingers, hair, and clothing from getting caught in the Münter hitch.

Operating (Tending) the Tandem Prusiks

The tandem Prusiks should be tended by a belayer wearing gloves (Figure 6-11).

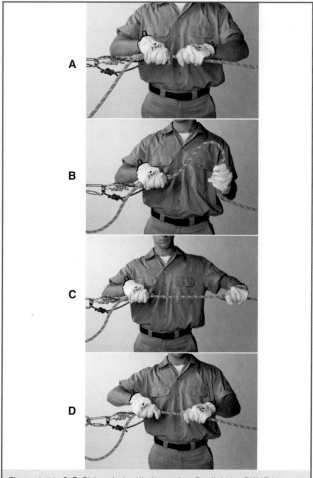

Figure 6-11 A-E, Giving slack with the tandem Prusik belay. **F-H,** Taking up slack.   *Continued*

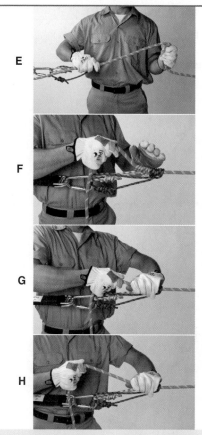

Figure 6-11 A-E, Giving slack with the tandem Prusik belay. **F-H,** Taking up slack.

▨ *Warning*

Belayers must be alert at all times. Failures often occur with no warning. While belaying with tandem Prusiks, do not allow the rope to come into contact with any part of your body other than your hands. You could be injured if the belay activates.

Prusik Minding Pulley (PMP)

Rigging of the Prusik minding pulley is shown in Figure 6-12.

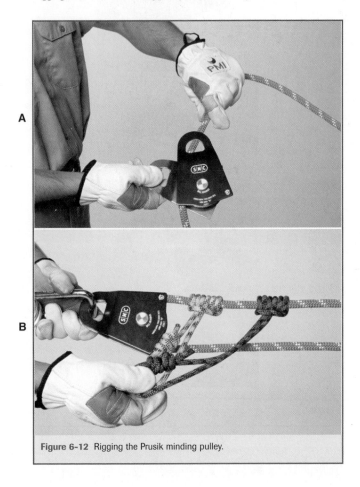

Figure 6-12 Rigging the Prusik minding pulley.

Pickoff Rescue

Equipment Needed for Rescue of a Person Wearing a Seat Harness

- One main-line rope with an adequate safety factor for a two-person load.
- One sewn, manufactured seat harness with thigh and leg supports for the rescuer.
- One rappel device with enough friction to handle the weight of two people and, preferably, with variable friction.
- Two large locking carabiners (in addition to the locking carabiner already in the rescuer's seat harness tie-in point).
- One short sling (approximately 2 feet [0.6 m]) with loop in both ends, or an adjustable rescue pickoff strap that will support one person's weight with an adequate safety factor.

Sizing Up the Situation in a Pickoff Rescue

Physical Situation

- Is the subject secure or is he or she in immediate danger of falling?
- Is there loose debris above the subject that your rope could dislodge?

Emotional Situation

- Is the subject hostile?
- Is the subject trying to grab you?
- Will the subject follow directions?
- Is the subject comfortable in the high angle environment?
- Is the subject experienced with high angle work so that he or she can assist in the procedure?

Initial Assessment of the Subject's Medical Needs

- Is the subject conscious and alert?
- Is the subject breathing?
- Is there uncontrolled bleeding?
- Is spinal injury a possibility?
- What are the obvious injuries?
- What are the subject's physical complaints?

Communicating with the Subject

- Reassure the individual.
- Tell the subject who you are.
- Tell the subject exactly what you plan to do.
- Ask if the subject is injured.
- Describe in detail how you will do the rescue.
- Tell the subject how he or she can help:

1. Do not move until told to do so.
2. Do not grab anything unless told to do so.

Pickoff Rescue of a Subject Not Wearing a Seat Harness

Placing a Tied Seat Harness on a Subject

Any of a number of tied harnesses might be placed on a subject for a pickoff rescue. Some considerations in deciding which to use include the following:

- The rescuer should be able to tie the harness with a minimum of physical disruption to the subject.
- The harness should be quick and easy to tie under difficult conditions.
- The harness should be self-adjusting.

▓ *Warning*

- ❏ Before you go over the edge in a rappel, particularly in a rescue, check for loose personal items or vertical gear. All items and gear must be secured so that they do not fall out of pockets, packs, or gear slings. In addition to injuring the rescue subject or other rescuers, you may lose an essential piece of equipment just when you need it the most.
- ❏ Always do a last minute safety check. Make sure harness buckles are correctly secured and that all carabiners are locked and aligned in the correct manner of function. Make sure knots are tied correctly and anchors are secure. Check for any loose clothing or hair that might be drawn into the descender. Make sure your helmet is secure.

Tied Hasty Seat Harness

The hasty seat harness is one type of seat harness that can be placed on a subject with a minimum of disruption. It is created from a length of tubular webbing 10 to 15 feet (3 to 4.6 m) long, depending on the size of the subject. The webbing is tied into a continuous loop using a ring bend (water knot) backed up before the rescuer begins the rappel.

Tying the Hasty Seat Harness (Figure 6-13)

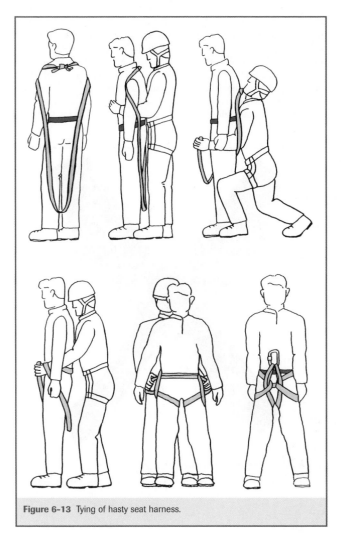

Figure 6-13 Tying of hasty seat harness.

Rescue Chest Harness

A rescue chest harness is used for subjects who have problems remaining upright in a seat harness, such as people with a large upper body in relation to the lower body or a "spare tire." They tend to lean back or even fall over backward when positioned only in a harness. Placing a chest harness on a rescue subject can help hold the person upright. Figure 6-14 shows a rescue chest harness. *The chest harness must not be used by itself in a pickoff rescue; it must be used* in combination with a rescue seat harness, such as the hasty seat harness. It may also be used with a sewn, manufactured seat harness to help hold a subject upright.

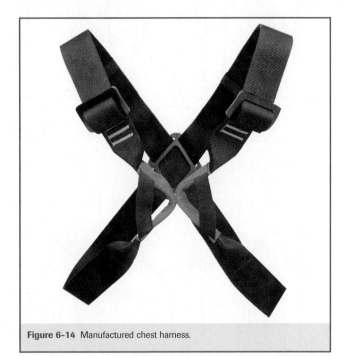

Figure 6-14 Manufactured chest harness.

Tying the Rescue Chest Harness

Figure 6-15 shows the tying of a quick chest harness that can be used with a seat harness for a pickoff rescue.

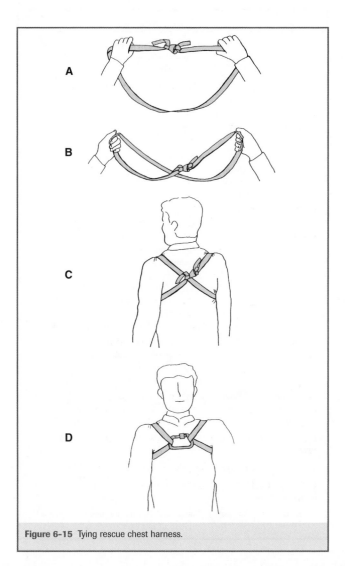

Figure 6-15 Tying rescue chest harness.

High Angle Lowering

Lowering Commands

Signal	Phrase Meaning
Litter tender to brakeman: *"On belay?"*	The litter team is ready. They should not proceed without a reply from the brakeman.
Brakeman to litter tender: *"On belay!"*	The braking system is engaged and ready to be loaded.
Litter tender to brakeman: *"Down slow"* or *"Down fast."**	The brakeman allows rope through the brakes slowly or faster, per the litter tender's request.
Litter tender to the brakeman (this signal may be given by anyone who sees danger or potential problems developing): *"Stop!"*	All rescuers stop activity.
Litter tender to brakeman (or vice versa): *"Stop! Stop! Why stop?"*	This signal is given when, for an unexplained reason and without command from the litter tender, the rope has stopped moving. The brakeman may still be letting out rope, but the rope may be jammed somewhere. This obviously has the potential for creating a very serious problem and requires an explanation.
Brakeman to litter tender: *"Zero!"*	Only about 20 feet (6 m) of rope is left. The litter team should set the litter down at a convenient spot so that a new brake and anchor set can be established.
Litter tender to brakeman (and belayer when used): *"Off belay."*	The litter, rescue subject, and litter tender or tenders are on the ground or in a secure position and in no danger of falling.
Brakeman to litter tender: *"Belay off."*	Acknowledges litter tender's signal.
The following voice communications may also be used when needed.	
Litter tender to brakeman or belayer: *"Slack."*	The rope is too taut; give us some slack.
Litter tender to brakeman or belayer: *"Tension."*	Take up some rope and make it more taut to help us out here.

Continued

Signal	Phrase Meaning
Litter tender to brakeman or belayer: *"Slack on belay line"* or *"Slack on main line."*	When a belay is used, the tender must specify which line he or she is talking about.
Litter tender to brakeman: *"Off rope."*	The litter has reached the ground, and the litter tenders have unclipped the rope from the litter and no longer need the line.

*NOTE: The "Down slow" or "Down fast" signal should be repeated by the brakeman back to the litter tender so that the tender knows the brakeman understands. Otherwise, the litter may be lowered at a rate different from the one desired.

Using a Brake Bar Rack to Lower a Practice Rescuer on a Slope

Figure 6-16 shows how to lace a brake bar rack for lowering.

Figure 6-16 Brake bar rack lowering on slope. *Continued*

Figure 6-16, cont'd

The Spider (also called a *Bridle* or *Harness*)

Creating an Adjustable Spider from Rope

Figure 6-17 shows an adjustable litter spider made from rope. The materials needed include:

■ Two lengths of static kernmantle rope, each 12 feet (3.6 m) long and at least ⅜ inch (9.5 mm) in diameter.

■ Four lengths of Prusik material, each 4 feet (1.2 m) long. The Prusik cord must be of appropriate diameter to grip the static kernmantle rope. Examples: 7 mm Prusik cord on a ⁷⁄₁₆-inch (11 mm) rope, and 8 mm on a ½-inch (12.5 mm) rope.

■ Four locking carabiners with gate openings large enough to fit over a litter rail.

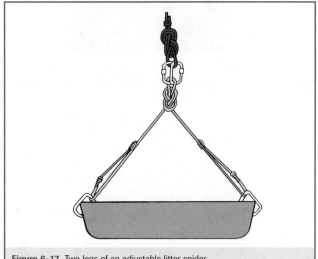

Figure 6-17 Two legs of an adjustable litter spider.

Litter Tender Tie-Ins

Figure 6-18 shows a litter tender tie-in system using two ascenders.

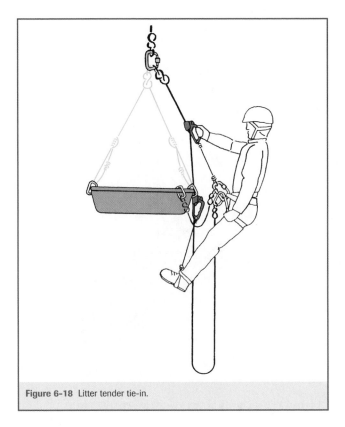

Figure 6-18 Litter tender tie-in.

Belay System for Litter Lowering

Figure 6-19 shows a belay system for litter lowering.

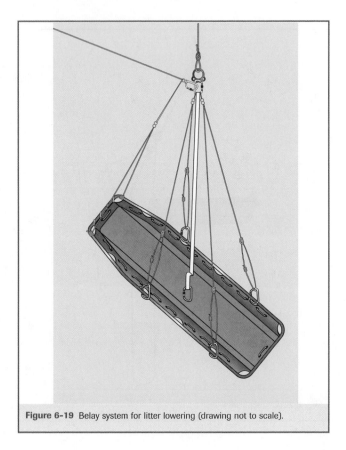

Figure 6-19 Belay system for litter lowering (drawing not to scale).

Rigging the Litter for Single Line Lowering

Figure 6-20 shows the rigging of a litter for single line lowering.

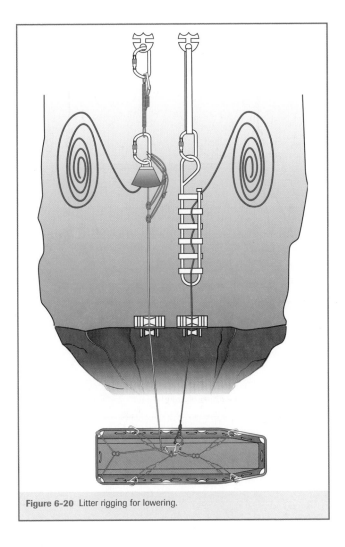

Figure 6-20 Litter rigging for lowering.

Spiders for Double Line Lowering

Figure 6-21 shows the spider system for a double line lowering.

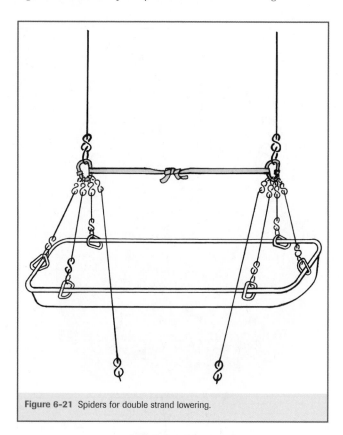

Figure 6-21 Spiders for double strand lowering.

Passing Knots

Figure 6-22 shows a procedure for passing knots. This system uses a single line lower with a belay (which, for clarity, is not shown). The following equipment is needed in addition to a regular lowering system:

■ Separate anchored braking system
■ Short length of rope (about 25 feet [7 to 8 m]) for interim lowering
■ Rope grab for each rope in the main lowering system

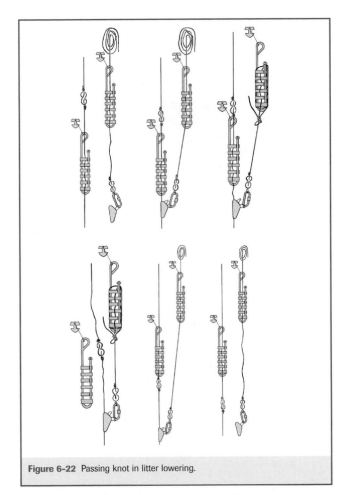

Figure 6-22 Passing knot in litter lowering.

Hauling Systems

1:1 Theoretical Mechanical Advantage (TMA) Hauling System

Figure 6-23 shows a 1:1 hauling system for raising a load up a vertical drop. The minimum equipment requirements for this system are:

- One main line rope
- One anchor sling
- Two locking carabiners
- One rope grab (progress capture device [PCD])
- One load-releasing hitch
- Separate belay systems appropriate for the load being hauled

Figure 6-23 1:1 hauling system.

2:1 Hauling System without the Diminishing V

Figure 6-24 shows one way to have a 2:1 TMA. Note that only a single haul line attachment is going to the load, as would be the case in a 1:1 hauling system. However, after the rope has gotten to the top, a short 2:1 hauling system is attached to it with a haul cam. Because the diminishing V is smaller and kept at the top, the system is less likely to snag, and if it does, the rescuers are able to reach it to free it.

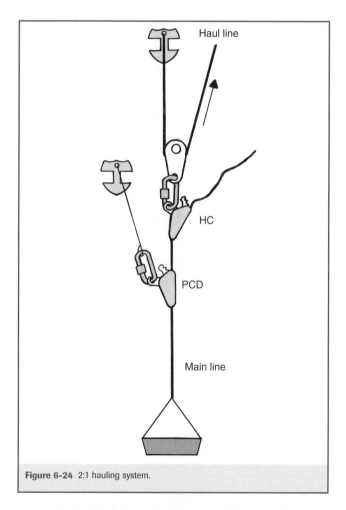

Figure 6-24 2:1 hauling system.

3:1 Haul System (z-Rig)

Figure 6-25 shows a 3:1 haul system. This particular 3:1 system is also commonly known as a *Z-rig* because of the approximate shape the rope makes as it goes through the system. The minimum equipment requirements for this system are:

- One main line rope
- Three locking carabiners
- Two pulleys
- Two rope grabs (one for hauling, one progress capture device [PCD])
- Separate belay systems appropriate for load being hauled

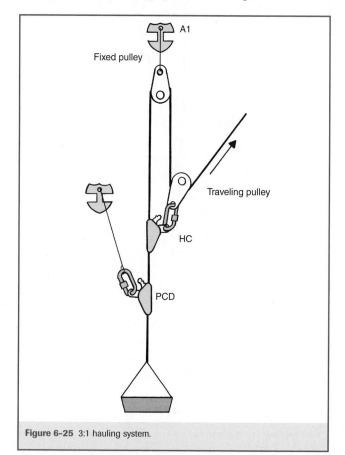

Figure 6-25 3:1 hauling system.

4:1 Hauling System (Piggyback System)

Figure 6-26 shows a 4:1 haul system, also known as a *piggyback system or pig-rig*. The minimum equipment requirements for this system are:

- One main line rope
- One hauling rope (50 to 100 feet [15 to 33 m], depending on the space available for the haul)
- Three locking carabiners
- Two pulleys
- Two rope grabs (one for hauling, one PCD)
- Separate belay systems appropriate for load being hauled

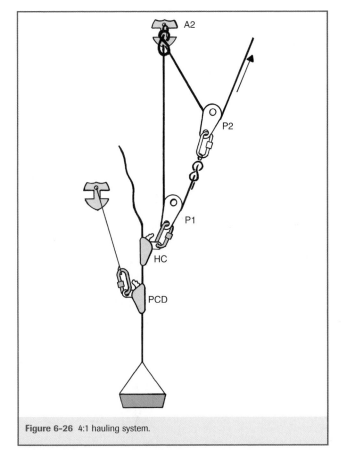

Figure 6-26 4:1 hauling system.

7

Rigging
Calculations

Calculating the Fall Factor

The fall factor is calculated by dividing the distance the person on the rope falls by the length of the rope between him or her and the point of protection:

$$\frac{\text{Distance person falls}}{\text{Length of rope}} = \text{Fall factor}$$

Warning

Although calculating the fall factor is useful for making general estimates of the forces on a rope in a fall, other factors often affect the situation, including the following:

❑ The rope may be running through intermediate points of protection, or it may be rubbing against rock or other surfaces. This creates drag that may slow the rope's ability to stretch. This in turn effectively shortens the length of rope available to absorb the energy of the fall. The net result would be a much higher fall factor and greater impact forces on the person and on parts of the system, such as the anchors.

❑ In most climbing situations, a fall does not occur completely in free air. The fall factor is of little importance if the person falling slams into the ground or is bashed against the wall on the way down.

❑ Recent tests have shown that the fall factor formula for dynamic ropes does not necessarily translate to static ropes, especially when very short or very long lengths of static rope are involved.

Tower Rescue Size Up

Determine the safety of the tower. If the tower is a guyed tower, look for loose, damaged, or missing guy wires. Check the integrity of the tower (i.e., look for any damage). If a winch or some other form of lifting device is on site, you must gain control of that equipment or perform a lockout/tagout on it.

Determine the number of rescue subjects. Once you know the number of subjects, you can allocate your resources accordingly.

Determine the location of the rescue subject or subjects. Their location on the tower may help in determining the number of rescuers needed on the tower. On a monopole, only a limited number of rescuers can reach the subject. A small face on a guyed tower doesn't afford the same room for working as a free standing tower might.

Determine as best you can the medical needs of the rescue subject or subjects. Are they breathing? Can you tell if they are conscious or unconscious? Are they uninjured, slightly injured but can help with the rescue operation, or injured and unable to help? This information can help you determine the number of rescuers needed on the tower.

Determine the type of rescue needed. Will it be a rescue or recovery? Will the subject need an attendant? Will this be a tower-based or a ground-based rescue?

Determine the number of rescuers needed and available. How many rescuers can you get into position on the tower safely? Do you have enough trained rescuers to perform the rescue from the tower?

Develop a rescue plan. Will it be a ground-controlled rescue or a tower-based rescue? Your rescue plan will be guided by the information you already have gathered. If the rescue is to be tower based, will the lowering be controlled from the tower or by the attendant? If it is to be ground based, are there antennas or dishes around which the rescuers must negotiate? If so, how will that be done?

Select your rescuers. Your rescuers must understand and agree to the proposed rescue plan, and they must be able to perform their assigned tasks. They should have previous tower training and experience and must be comfortable at the height at which they will be working.

Determine your climbing system. Will you use the existing fall protection on the tower? Will you perform a lead climb or a lanyard climb? What are the advantages or disadvantages of the system you have chosen, and do your rescuers understand them?

Consider both the current and predicted weather. As the weather changes, so will your safety considerations. High winds may require an attendant, even if the original plan did not call for one.

Consider the time of day. If the rescue extends into the night, are the rescuers prepared for cooler or colder weather? Do they have two sources of hands-free lighting available to them? Does the rescue plan take into consideration that the rescue may take longer because of darkness? Is an attendant now needed?

Develop a backup plan. In case your original plan won't work, have you considered some alternatives? If you have a limited number of qualified tower rescuers and one of them should require help, what is your backup plan?

10% Rule for Highlines

The 10% rule is a conservative method of tensioning a highline. According to the 10% rule, the center of the unloaded highline should sag a vertical distance of about 10% of the span for every 200 pounds (90 kg) of expected load and every 100 feet (30 m) of span in the rope. Unless a rescuer incorporates some sort of force measurement system and is thoroughly experienced in the effects of their rigging, a conservative approach is recommended.

REMEMBER: When calculating the 10% rule, the calculations should be based on the *total weight* of the load, not just the weight of the individuals on the load. The calculations must also consider the *total length* of the span between the two supports, such as the anchors, not just the width of the gap bridged.

Rope sag is the visible amount of sag in the main line before the load is applied. There are two variables in the 10% rule. If either of them changes, the amount of sag must change.

Examples:

For a 200-pound (90 kg) load *(1L)* on a 100-foot (30 m) span, the formula is:

$$1L \ [90 \ kg] \times 100 \ [30 \ m] \times 0.10 = 10\text{-foot} \ [3 \ m] \ \text{sag required}$$

For a 200-pound (90 kg) load *(1L)* on a 200-foot (60 m) span, the formula is:

$$1L \times 200 \ (60 \ m) \times 0.10 = 20\text{-foot} \ (6 \ m) \ \text{sag required}$$

For a 400-pound (180 kg) load *(2L)* on a 200-foot (60 m) span, the formula is:

$$2L \times 200 \ (61 \ m) \times 0.10 = 40\text{-foot} \ (12 \ m) \ \text{sag required}$$

8

Helicopters

Helicopters: Decision Making and Situational Awareness

Before launching a helicopter for a rescue or requesting the assistance of an outside agency with aviation assets, make sure you are not "wrapped around the axle" by the excitement of the moment. First, ask yourself the following three questions:

Is this plan in the best interest of the subject and of rescuer safety?

Is there a better way to do it?

If darkness or poor weather is hampering a helicopter response, will the subject be better served by an evacuation in the morning or in improved conditions?

Additional questions to ask yourself when deciding whether to conduct a helicopter rescue include the following:

- Is a safe landing site available within reasonable distance of the accident site?
- Does the urgency of the rescue subject's condition require getting someone to the accident site as quickly as possible?
- Would immediate insertion of a rescuer who can provide advanced life support (ALS) care convert an urgent medical case to a lower priority ground evacuation?
- Is the risk associated with traversing terrain to the incident site greater than the risks posed by specialized helicopter rappel, short haul, or hoisting techniques?
- Are all the rescuers proficient with the helicopter rescue technique under consideration?
- Do extreme environmental factors prevent the use of a helicopter?
- Is communication with all involved rescue personnel adequate, or do communications barriers exist?

Emergency Briefing Format*

Here's what I think we face.

Here's what I think we should do.

Here's why.

Here's what we should keep our eye on.

Now, talk to me.

Helicopter Landing Zones

- *Touchdown pad:* A designated area, which may involve a prepared or improved surface, where the helicopter skids are in contact with the ground.

---

*Modified from Weick K: South Canyon revisited: lessons learned from high reliability organizations, Paper presented at Decision Workshop on Improving Wildland Firefighter Performance Under Stressful, Risky Conditions: Toward Better Decisions on the Fireline and More Resilient Organizations, Missoula, Montana, June 12-16, 1995.

- *Safety circle:* An obstruction-free area on all sides of the helicopter touchdown pad (Figure 8-1) that permits safe approach and departure at the touchdown pad. It must be clear of obstacles and hazards (e.g., wires, tall trees, loose debris) for a helicopter departing at a low angle (Table 8-1).
- *Other Considerations.* An exposed ridgeline, outcropping, or clear saddle can provide a landing zone, especially if the approach and departure paths

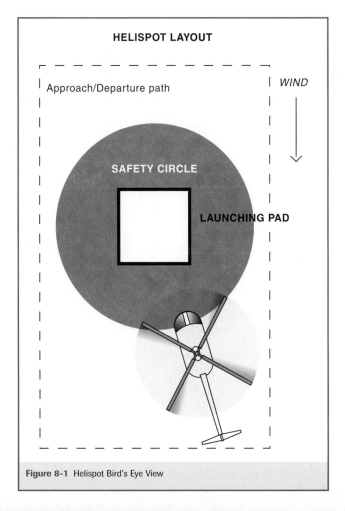

Figure 8-1 Helispot Bird's Eye View

Table 8-1	Helicopter Touchdown Pad and Safety Circle Dimensions	
Helicopter Type	**Touchdown Pad Dimensions**	**Safety Circle Diameter**
Type 1 (heavy helicopter)	30 feet × 30 feet (9 m × 9 m)	110 feet (33.528 m)
Type 2 (medium helicopter)	20 feet × 20 feet (6 m × 6 m)	90 feet (27.432 m)
Type 3 (light helicopter)	15 feet × 15 feet (4.6 m × 4.6 m)	75 feet (22.86 m)

From the Department of the Interior, Bureau of Land Management: *Interagency helicopter operations guide (IHOG)*, National Interagency Fire Center (NIFC), NFES Item #1885, Boise, Idaho, 2002.

are unobstructed (Figure 8-2). However, take into account that winds can be variable. In a deep canyon, a helicopter requires a long forward and unobstructed path to gain altitude. A tight, confined landing area requires a maximum performance takeoff. This may approach the operating limits of the aircraft and should be avoided.

- *Slopes.* Avoid slopes over 5 degrees or 11% grade. On sloping surfaces, a situation known as *dynamic rollover* can occur. When a helicopter is lifting off from a sloping landing zone, if one skid or wheel is still on the ground, the helicopter may end up pivoting around this skid or wheel. When this happens, the pilot can reach the "stop" limit on the cyclic control as cyclic travel runs out, and he or she will not be able to stop the rollover. If the aircraft is allowed to continue to roll, a critical angle may be reached at which the roll cannot be corrected, and the helicopter will roll over onto its side. Dynamic rollover also occurs on level surfaces, as when a skid or wheel becomes stuck on a landing zone (e.g., sinks in mud or warm asphalt) or is restrained by a tie-down or some other obstacle during takeoff.
- *Dust abatement.* For repeated use of a landing zone, especially an unpaved helibase, consider dust abatement, using available fire apparatus to wet down the surface. Reducing dusty conditions improves safety, because the pilot can observe the touchdown pad during landing and receive unobstructed hand signals from ground personnel. Furthermore, dust and sand can damage aircraft engines.
- *Wind indicator.* A wind indicator provides the pilot with a visual cue of wind direction and speed. A wind indicator can be a windsock, flagging, or smoke or dirt thrown into the air as the helicopter initiates a high orbit

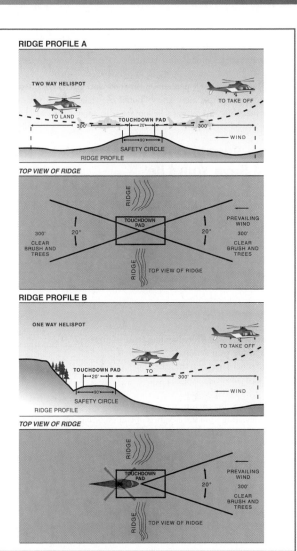

Figure 8-2 Helispot Profile View

over the scene. A helicopter achieves optimum performance when landings and takeoffs are made into the direction of the oncoming wind. Ground rescuers should anticipate this in their selection of a helispot and in the staging of rescue apparatus or personal equipment.

- *Snow landing*. Depth perception on snow and glacial ice often is poor. To reduce blowing snow, snow should be packed down with skis or a snow machine. Equipping a helicopter with snow pads can help reduce uncontrolled settling of the aircraft in snow.

Twelve Standard Aviation Questions That Could Save Your Life

Is this flight necessary?

Who is in charge?

Are all hazards identified and have you made them known?

Should you stop the operation or flight due to:

- Inadequate and unclear communications
- Hazardous weather
- Turbulence
- Insufficient or untrained personnel
- Conflicting priorities
- Deceased rescue subject

Is there a better way to do it?

Are you driven by an overwhelming sense of urgency?

Can you justify your actions?

Are there other aircraft in the area?

Do you have an escape route?

Are any rules being broken?

Are communications getting tense?

Are you deviating from the assigned operation or flight?

WHEN IN DOUBT—DON'T!

Operational Red Flags

- Using media aircraft.
- Conducting a rescue with an unknown crew or aircraft.
- Exceeding the operating capabilities of the aircraft or crew.
- Improvising with an unpracticed or recognized technique.
- Radio incompatibility problems (e.g., VHF versus UHF).
- Preoccupation with minor details—not maintaining the "big picture."
- Inadequate leadership, failure to designate command.
- "Pressing"—mission-itis dictates operational decision making.
- Failure to delegate tasks and assign responsibilities.
- Failure to communicate intent and plans (i.e., lack of a briefing).

Index